With this book,
you have helped
plant a tree.
THANK YOU

For my children, that they may never forget that I love them
and will continue to fight for the future they deserve.
L. D.

DUST & ASHES
PUBLISHING COMPANY

ISBN: 978-1-7360455-2-7 (Paperback)
ISBN: 978-1-7360455-5-8 (Hardcover)
ISBN: 978-1-7360455-3-4 (eBook)

Library of Congress Control Number: 2021924336

Illustrations by Iva Pažin.
Book design by Lisette M. Díaz Aponte.

First printing edition 2021.

www.lisettediaz.com
ldiaz@dapublishingco.com

FOR THE LOVE OF TREES

WRITTEN BY
LISETTE M. DIAZ

ILLUSTRATED BY
IVA PAŽIN

I am a child,
and I love trees.

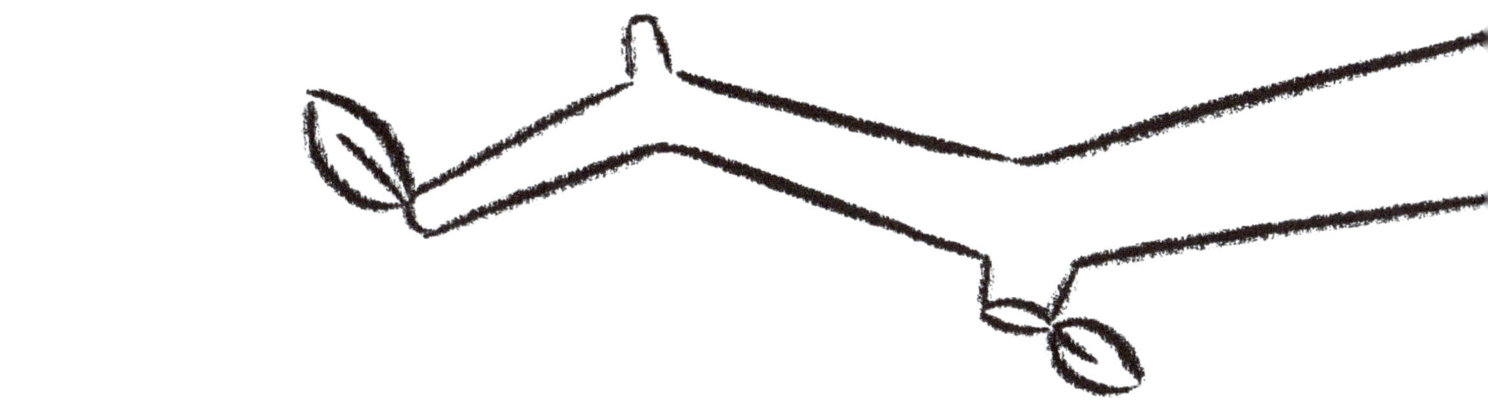

Each day,
I climb their branches
to get some exercise.

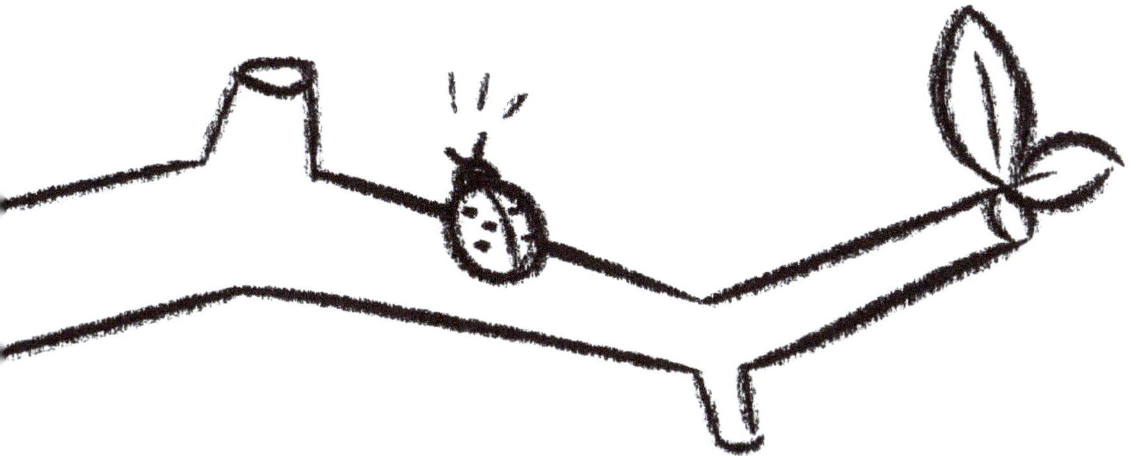

When it gets too hot,
I sit under their shade.

When I am hungry,
I eat their fruit.

When I am sad,
their trunks are easy
to hold.

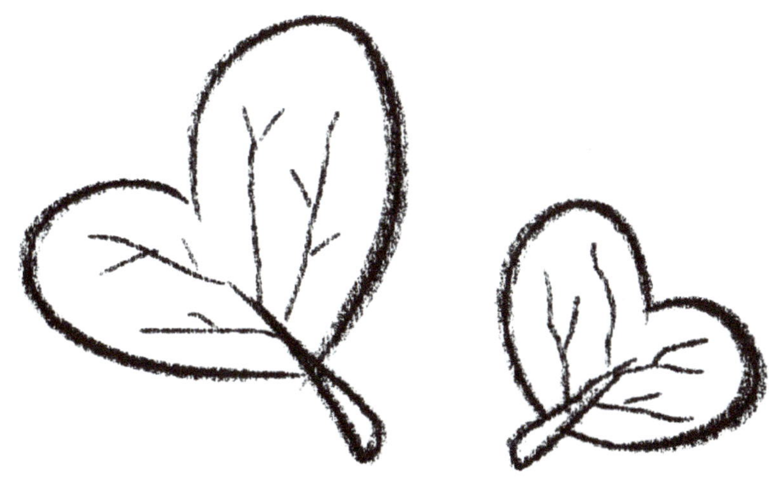

When I am curious,
I watch the animals
scurry all over them.

When my back is itchy, trees make great scratchers.

I think the bears agree.

When I want adventure,
they make wonderful
pretend places, like...

...pirate ships...

...rockets...

...towers...

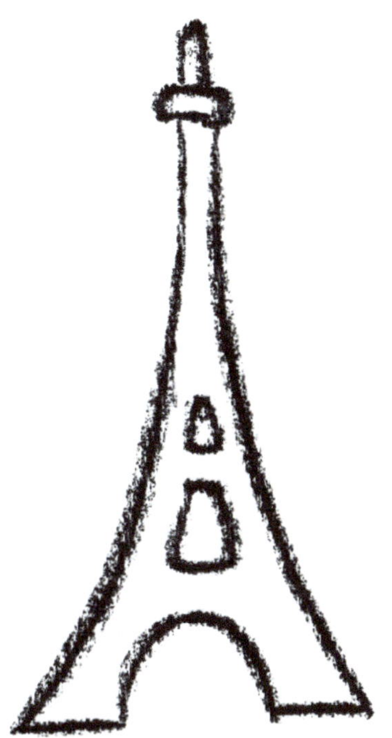

...and castles.

Out on busy roads
I find it hard to breathe.

But when I am among
my trees, I can take
deep breaths of fresh air.

But the trees are disappearing.

There used to be so many. Now, just a few are left.

The animals have
fewer homes.

Our air is not
very clean.

So, I'm helping all the animals.
I'm helping our earth breathe.
I'm getting my hands dirty.
I'm helping to grow trees.

If we work together
it will be a breeze.
Won't you join in my adventure?
Let's go plant some trees!

We are looking forward
to a better tomorrow.

www.ingramcontent.com/pod-product-compliance
Lightning Source LLC
Chambersburg PA
CBHW042154030726
47599CB00004B/729